CW00553567

The Genetic Code

Decoding the Blueprint of Life

CLARA NEWTON

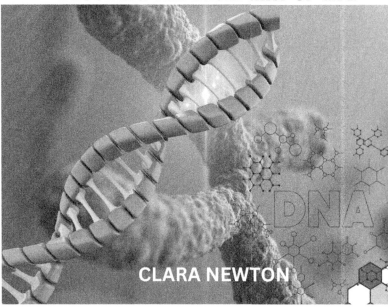

THE GENETICS CODE

CODE

DECODING THE BLUEPRINT OF LIFE

CLARA NEWTON

COPYRIGHT

Book Description:

The complete guide to the intriguing field of genetics is "The Genetic Code: Decoding the Blueprint of Life." This essential book examines the fascinating history of genetics, its significance in contemporary culture, and its potential to fundamentally alter our way of life.

Readers are given a general overview of the significance of genetics in contemporary culture in Chapter 1, Introduction. The chapter gives an overview of the book and its structure while highlighting some of the most significant individuals in the history of genetics.

Readers will get a thorough knowledge of the fundamental mechanics of heredity, how features are handed down from one generation to the next, and how genetic diversity develops, from Chapter 2, The mechanics of heredity. Mendelian genetics, meiosis, and sex-linked characteristics are only a few of the core concepts covered in this chapter.

The discovery of DNA, its structure, and the replication and transcription processes are covered in Chapter 3: The Structure and Function of DNA. This chapter also examines how DNA functions as a genetic information carrier and how the variety of life on Earth is a result of it.

The many forms of genetic mutations and how they might result in illnesses and disorders are covered in Chapter 4, Genetic Mutations and Disease. The chapter also covers gene therapy's potential to treat or perhaps reverse certain diseases and presents the most recent findings in this field.

Chapter 5 of this book, Genomics and Individualized Therapy, walks readers through the Human Genome Project, genome sequencing and analysis, and the promise of individualized therapy based on unique genetic profiles. Additionally covered in this chapter are the ethical and legal ramifications of genome sequencing as well as the prospective advantages and disadvantages of customized treatment.

The ethical and sociological ramifications of genetic alteration, including the genetic manipulation of plants and animals as well as the use of CRISPR technology, are the main subject of Chapter 6 on Genetic Engineering and Biotechnology. The chapter also looks at the possible uses of genetic engineering for developing new medications for illnesses as well as any negative effects that might result from these developments.

Human evolution and the genetic foundation of variation are only two of the topics covered in Chapter 7's discussion of evolutionary genetics. The chapter also discusses the most recent findings on genetic variation, speciation, and their implications for our comprehension of nature.

Readers are exposed to the larger social consequences of genetic research, such as the possibility of genetic testing and the possibility of genetic discrimination, in Chapter 8, Ethical and Social Consequences of Genetic Research. The chapter also looks at how genetics fits into society and how genetic research could aggravate already-existing disparities.

Finally, readers are encouraged to consider the future of genetics and how it could change our lives in Chapter 9, Conclusion. The chapter examines some of the most recent genetic research breakthroughs and their prospective uses in the medical industry, agriculture, and other industries. It also covers the difficulties and possibilities that this fascinating area of study may face in the future.

The book "The Genetic Code: Decoding the Blueprint of Life" offers a thorough analysis of this rapidly developing discipline and its effects on our daily lives. This book is a must-read for anybody interested in genetics and the implications it has for the future due to its straightforward explanations, interesting writing style, and thought-provoking material.

Table of Contents:

Chapter 1:

Introduction

Thank you for visiting "The Genetic Code: Decoding the Blueprint of Life." The subject of genetics is quickly developing and altering how we view life. Genetics is a key component in understanding the nature and behavior of living things, from the smallest bacterium to the greatest animals. With new discoveries and scientific advancements occurring every day, it is a discipline that is continually pushing the limits of our understanding.

From the earliest days of Mendelian genetics to the most recent developments in CRISPR technology, we examine the history of genetics in great detail in this book. We examine the ways that genetic mutations can cause disease, the mechanics of heredity, the structure and operation of DNA, and more. We also take into account the social and ethical ramifications of genetic research, such as the use of genetic testing to identify illness risk or the possibility of using genetic engineering to change the trajectory of human evolution.

We make an effort to give a fair and instructive review of the area of genetics throughout the whole book. This book provides a fascinating and thought-provoking look into an area that is transforming our knowledge of life itself, whether you are a scientist, student, or simply a curious reader. Join us as we travel through "The Genetic Code: Decoding the Blueprint of Life" so that you may learn more about life.

The Importance of Genetics

A key area of research that is vital to our comprehension of life itself is genetics. It includes research on heredity, variation, and the molecular processes that control the composition and operation of biological things. Numerous fields, including health, agriculture, evolutionary biology, and conservation, highlight the significance of genetics.

The discipline of medicine is one of genetics' most significant applications. It is now possible to diagnose and treat genetic disorders sooner thanks to the discovery of several disease-causing genes through genetic research. For instance, the identification of the cystic fibrosis-causing CFTR gene has sparked the creation of medications that specifically target the underlying genetic flaw. Similar to this, the discovery of the BRCA1 and BRCA2 genes—linked to a higher chance of developing breast cancer—has prompted the creation of screening programs and preventive measures for those who are at risk. The development of gene therapy, which has the potential to treat genetic illnesses by replacing or correcting defective genes, is another result of genetic research.

The study of genetics is significant in agriculture. With the use of selective breeding, it is now possible to create crops and livestock with desired characteristics, including increased yields, enhanced disease resistance, and greater nutritional value. A reduction in the use of hazardous chemicals in agriculture is possible thanks to the creation of crops that are resistant to pests and herbicides thanks to genetic modification. These improvements in agricultural genetics have the potential to reduce hunger and increase agriculture's sustainability.

Genetics is crucial to our knowledge of evolutionary biology and conservation, in addition to its uses in health and agriculture. We can trace the development of populations and comprehend the principles behind evolution by studying genetic variation both within and across species. Thanks to genetic research, we can now recognize populations that are in danger of becoming extinct and create conservation plans to save them.

Furthermore, forensic science heavily relies on genetics. Criminal investigations use DNA testing to locate suspects, prove paternity, and connect people to crime sites. The use of DNA fingerprinting as a routine investigative method has helped to solve several crimes and acquit innocent people.

Overall, the study of genetics is essential to our comprehension of the natural world and has several applications in forensic science, health, agriculture, and conservation. The study of genetics has the potential to boost food security, save endangered animals, prevent crime, and improve human health. As a result, it is a field of study that is still very important to scientists, politicians, and the general public.

Key Figures in the History of Genetics

The history of genetics spans decades of research and development in science. Many influential people have significantly contributed to our knowledge of genetics and its applications throughout the years. Listed below are some of the most significant individuals in genetic history:

Eugene Mendel

For his groundbreaking work on pea plants in the middle of the 1800s, Gregor Mendel earned the title "father of genetics." His research helped to uncover the fundamental rules of inheritance, such as the laws of segregation and independent assortment. Mendel's research established the field of contemporary genetics and offered a conceptual framework for comprehending the transmission of hereditary features from one generation to the next.

Morgan and Thomas Hunt

In the early 1900s, Thomas Hunt Morgan was a well-known geneticist best known for his research on fruit flies. His research led to the identification of sex-related features, which are present on both the X and Y chromosomes. Genetic recombination, which happens when genetic material is transferred between chromosomes during meiosis, was also discovered by Morgan. His research contributed to the development of the chromosomal theory of inheritance, which contends that genes are found on chromosomes and are transmitted from parents to children in a known manner.

Franklin Rosalind
Early in the 1950s, British biophysicist Rosalind Franklin made fundamental contributions to the understanding of DNA structure. Her X-ray crystallography research contributed to the creation of the earliest pictures of the double helix structure of DNA. Her work was crucial to James Watson and Francis Crick's discovery of the structure of DNA, which led to their eventual recognition with the 1962 Nobel Prize in Physiology or Medicine.

Francis Crick and James Watson
The geneticists James Watson and Francis Crick, who discovered the structure of DNA in 1953, are perhaps the most well-known in history. They proposed the double helix structure of DNA using X-ray crystallographic data from Rosalind Franklin and Maurice Wilkins. By establishing a

molecular foundation for comprehending how genes are copied and transmitted from generation to generation, this discovery transformed the discipline of genetics.

McClintock, Barbara

American geneticist Barbara McClintock made important contributions to the study of chromosomes and their activity via the science of cytogenetics. She discovered transposable elements, or "jumping genes," which may migrate from one spot on a chromosome, as a result of her studies on maize plants in the 1940s. McClintock's research shattered the conventional wisdom that genes were immobile in their positions on chromosomes and advanced knowledge of genetic control and diversity.

Venter, Craig

Modern scientist Craig Venter is well recognized for his 1990s work on the Human Genome Project. He was in charge of the privately financed project to sequence the human genome, a challenging endeavor that entailed cataloging and mapping every gene in the genome. This research has opened up fresh perspectives on the genetic causes of illness and helped establish the foundation for customized medicine.

Jillian Doudna

A molecular researcher named Jennifer Doudna is most recognized for developing the CRISPR-Cas9 system, a ground-breaking tool for genetic modification. Her work has helped to create a new generation of gene-editing tools that are quicker, more affordable, and more accurate than earlier ones. Through the ability to precisely modify genetic material, these techniques have the potential to change both agriculture and health.

In conclusion, the study of genetics has a long history, and many influential people have contributed much to our knowledge of genetics and its uses. The contributions of these individuals, from Mendel's research on pea plants through Doudna's work on CRISPR,

An Overview of the Book

A thorough investigation of the topic of genetics, "The Genetic Code: Decoding the Blueprint of Life" focuses on the discipline's history and influential personalities. The book gives a thorough summary of the most significant genetic advancements and discoveries, illustrating how they have changed our perception of genetics and its practical applications.

The book starts out by introducing readers to the history and discipline of genetics. The article talks about the early findings in the area, such as Gregor Mendel's research on pea plants, which established the rules of heredity, and Thomas Hunt Morgan's studies on fruit flies, which revealed the existence of genes on chromosomes.

The discussion of the discovery of DNA's structure, largely regarded as one of history's most important scientific breakthroughs, follows. The contributions of Rosalind Franklin, whose work on X-ray crystallography helped develop the first photographs of DNA's structure, are covered in depth together with the work of James Watson and Francis Crick in finding the double helix structure of DNA.

The book also discusses the identification of inherited illnesses, including sickle cell disease and Huntington's disease. It covers the use of genetic testing in modern medicine as well as how genetics plays a part in assessing illness risk.

The focus of the book then shifts to the present, where it discusses the most recent advances in genetics research and how they may affect the course of modern medicine, agriculture, and society as a whole. It talks about the continuous initiatives to map the human genome and how this has improved knowledge of genetics and illness.

The CRISPR-Cas9 gene-editing technique is one of the most fascinating advancements in genetics in recent years. The book provides an explanation of the background, growth, and prospective applications of this technology in both agriculture and medicine.

Readers learn about the most significant individuals in the history of genetics throughout the book, including Gregor Mendel, Thomas Hunt

Morgan, James Watson, Francis Crick, Rosalind Franklin, and Barbara McClintock. The book talks about their findings, their contributions to genetics, and how their ideas have shaped how we think about genetics today.

A broad range of readers, including students, researchers, and general readers who are curious about genetics and its significance, will find the book to be approachable and interesting. The book has several diagrams, pictures, and examples that serve to clarify the ideas covered in the text.

The book "The Genetic Code: Decoding the Blueprint of Life" provides a thorough and fascinating account of the field's history, its leading personalities, as well as the most recent advancements and their potential to influence the future. Anyone interested in genetics and its significance for science, medicine, and society at large should read it.

Chapter 2:

The Mechanisms of Inheritance

The procedures by which characteristics are handed down from one generation to the next are referred to as the "mechanics of heredity." This is accomplished via genetic information being passed from parent to child in the form of DNA (deoxyribonucleic acid).

Understanding how physical and behavioral qualities are handed down within populations and for the diagnosis and treatment of genetic illnesses require knowledge of inheritance. The mechanics of inheritance may be quite intricate, encompassing a wide range of genetic and environmental variables that affect how certain features manifest in various people.

From Gregor Mendel's early research on pea plants to more recent developments in genetic engineering and gene editing, several scientists have made substantial contributions throughout time to our knowledge of heredity. In an effort to understand the intricate interactions between genes, proteins, and other cellular processes that control the manifestation of inherited characteristics, scientists are still investigating the molecular mechanisms of inheritance.

Mendelian Genetics

Mendelian genetics is the study of the patterns of inheritance of characteristics from one generation to the next and is named after its originator, Gregor Mendel. The foundation of contemporary genetics is Mendelian genetics, which has aided scientists' understanding of the inheritance of physical and behavioral features as well as the transfer of genetic information.

Scientist and Augustinian monk Gregor Mendel experimented with pea plants at his monastery garden in Brno, Czech Republic, in the middle of the 19th century. In order to explore the inheritance of characteristics, Mendel crossed several strains of pea plants and then observed the progeny.

Mendel's research on pea plants resulted in the identification of the basic principles of heredity, commonly referred to as "Mendel's laws." The law of dominance, the law of independent assortment, and the rule of segregation are the three laws of inheritance.

According to the rule of segregation, each person has two alleles (gene forms) for each characteristic, and these alleles separate during the development of gametes (sex cells). This indicates that there is just one allele per characteristic present in each gamete.

According to the law of independent assortment, the inheritance of one trait is unrelated to the inheritance of other qualities. According to this, the segregation of alleles for one characteristic does not affect the segregation of alleles for another trait.

According to the rule of dominance, certain alleles are dominant and others are recessive. Recessive alleles can only be expressed when a person inherits two copies of the recessive gene; dominant alleles are always expressed.

Mendel's principles provide an elegant and straightforward explanation for how characteristics are passed down across families, but they are only relevant for traits that are controlled by a single gene with two alleles. In actuality, numerous genes and multiple alleles control the majority of features, and environmental circumstances often have an impact on how these traits are inherited.

Mendelian genetics is still a valuable resource for understanding the fundamentals of heredity and determining the possibility that a characteristic will be handed down from one generation to the next. The identification and management of genetic problems is one of the main applications of Mendelian genetics.

A person's genes' DNA sequence may change or become aberrant, which results in genetic diseases. Others are more complicated and are impacted by several genes and environmental variables. Some genetic abnormalities are inherited in a straightforward Mendelian fashion.

Based on a person's family history and the known inheritance patterns of the condition, mendelian genetics may be used to forecast the risk that they will acquire a genetic disorder. For instance, if an inherited disease is known to follow an autosomal recessive pattern, a person with one afflicted parent has a 50% chance of acquiring the condition.

Genetic counseling, which is the act of giving advice and assistance to people and families afflicted by genetic illnesses, also makes use of mendelian genetics. To assist people in understanding their risk of inheriting a genetic condition and in making educated choices about their reproductive options, genetic counselors use a range of resources, including family history, medical diagnostics, and genetic screening.

Mendelian genetics has been used to increase agricultural yields and animal breeding in addition to diagnosing and treating genetic problems. Farmers and breeders may increase the quality and output of their crops and livestock by selecting breeding plants and animals with desired features.

Overall, Mendelian genetics has had a significant influence on how we see heredity and has helped pave the way for several significant developments in genetics and medicine. Mendel's rules continue to be a valuable resource for forecasting the possibility of characteristics passing down from one generation to the next and for comprehending the fundamentals of heredity, even if they are not relevant to all traits and inheritance patterns.

Meiosis and Chromosome Segregation

Meiosis is a specialized kind of cell division that happens in sexually reproducing animals to convert diploid cells into haploid gametes (sex cells). Meiosis is composed of meiosis I and meiosis II, two rounds of cell division that are preceded by one cycle of DNA replication.

Meiosis' primary goal is to cut the number of chromosomes in half, ensuring that the developing zygote has the appropriate amount of chromosomes for the species when the haploid gametes unite during fertilization. By rearranging the genetic material via a process known as crossing over, meiosis also brings genetic variety into the population.

Meiosis is broken down into various phases, each of which is distinguished by certain occurrences and modifications in chromosomal structure and function.
During prophase I, chromosomes condense and couple together to form homologous pairs. Two sister chromatids, which are exact duplicates of

the same chromosome, make up each pair. During this phase, a process known as crossing over takes place, including the transfer of genetic material between homologous chromosomes.

Metaphase I: The metaphase plate, a plane that splits the cell in two, is where the homologous pairs align during metaphase I. Each homologous pair's orientation is unpredictable, which boosts genetic variety.
Homologous couples split and migrate to the cell's opposing poles during anaphase I. Each homologous pair sends one member to each daughter cell.

Telophase I and Cytokinesis: During telophase I, the chromosomes decondense as they approach the opposing poles of the cell and surround each pair of chromosomes with a nuclear envelope. Then cytokinesis takes place, producing two haploid daughter cells.

Meiosis II: Meiosis II resembles mitosis, except the offspring cells are haploid rather than diploid. The two meiotic cycles produce four haploid daughter cells, each of which has a different set of chromosomes because of the random distribution of homologous pairings and cross-overs.

The formation of gametes with the right chromosomal number depends on chromosome segregation during meiosis. Aneuploidy, a disorder in which cells have an aberrant number of chromosomes, may result from errors in chromosomal segregation.

Nondisjunction is a typical chromosomal segregation problem that happens when sister chromatids or homologous pairings are unable to correctly separate during meiosis I or meiosis II, respectively. As a consequence, one daughter cell gains an additional chromosome, while the other daughter cell loses a chromosome.

Genetic diseases like Down syndrome, which is brought on by an extra copy of chromosome 21, may result from nondisjunction. A single X chromosome in females causes Turner syndrome, and an extra X chromosome in men causes Klinefelter syndrome—two further instances of genetic abnormalities brought on by nondisjunction.

A sophisticated network of proteins and enzymes controls chromosomal segregation during meiosis to maintain appropriate chromosome alignment, separation, and distribution. The spindle apparatus, a collection of microtubules that connect to the chromosomes and pull them apart during cell division, is a crucial component of this process.

For sexual reproduction and genetic variation, meiosis and chromosomal segregation are crucial processes. Developmental defects and genetic illnesses may result from errors in these systems. To better comprehend these processes and create novel treatments for genetic illnesses, we must understand the mechanics of meiosis and chromosomal segregation.

Sex-Linked Traits

The X and Y chromosomes, which make up humans' sex chromosomes, are responsible for X- and Y-linked features. Compared to the Y chromosome, the X chromosome is bigger and has more genes. Because of this, the majority of sex-related features are connected to the X chromosome, whereas relatively few are linked to the Y chromosome.

Characteristics that are carried on autosomes, the non-sex chromosomes, as opposed to characteristics that are connected to sex, are inherited in various ways. Due to the fact that they have two copies of each autosome, boys and females generally have an equal probability of acquiring autosomal features. However, compared to men, who have one X and one Y chromosome, females have a distinct pattern of inheritance for X-linked characteristics since they have two copies of the X chromosome.

The inheritance of X-linked characteristics in females is the same as that of autosomal traits. A girl will still have a healthy copy of the gene on her other X chromosome even if she receives a recessive allele for an X-linked characteristic from one father. She won't display the characteristic as a consequence, but she will be a carrier and may pass it on to her progeny.

However, the inheritance of X-linked characteristics differs in men. Males only have one copy of each X-linked gene since they have one X chromosome. Accordingly, a man who receives a recessive allele for an X-linked characteristic from his mother will not have a second copy of the gene to mitigate the consequences of the recessive allele. He will display the characteristic as a consequence, even if the allele is recessive.

Color blindness is the most well-known instance of a sex-linked characteristic. A mutation in a gene on the X chromosome that codes for a protein important for color vision results in color blindness. Males are considerably more likely than females to have color blindness since the gene is on the X chromosome. Females may have the characteristic, but since they have a second, healthy copy of the gene on their other X chromosome, they are less likely to display it.

Hemophilia, a bleeding illness brought on by changes in genes on the X chromosome that code for clotting proteins, is another example of a sex-linked feature. Due to the fact that males have only one X chromosome and are thus more likely to receive a mutant gene, hemophilia is also more prevalent in men than in women.

In addition to hemophilia and color blindness, other sex-related traits include X-linked hypophosphatemia, which affects bone development, and Duchenne muscular dystrophy, a disease that results in muscle wasting and is brought on by mutations in an X-chromosome gene.

Pedigrees, which are diagrams that depict the inheritance patterns of certain features in a family, may be used to study sex-related attributes. Using pedigrees, one may identify whether a characteristic is sex-linked or autosomal and track its inheritance across many family generations.

Overall, since sex-linked characteristics are carried on the sex chromosomes, they are inherited differently from autosomal traits. Because men only have one X chromosome and are more likely to receive recessive genes, they are more likely to exhibit X-linked features. The inheritance of X-linked characteristics in females is the same as that of autosomal traits; however, females might carry X-linked traits without displaying them. Pedigrees may be used to study sex-related characteristics, identifying carriers and estimating the chance of the trait being passed down to the next generation.

Chapter 3:

The Structure and Function of DNA

Deoxyribonucleic acid, often known as DNA, is a sophisticated molecule that houses genetic material and is necessary for the growth, development, and overall health of all living things. DNA is made up of two strands of nucleotides that are wrapped around one another to form a double helix.

Deoxyribose, a kind of sugar molecule, a phosphate group, and a nitrogenous base make up each nucleotide. In DNA, nitrogenous bases come in four different varieties: adenine (A), thymine (T), guanine (G), and cytosine (C). A pairs with T and G pairs with C, respectively, forming particular pairings between the bases. The replication and transcription of DNA depend on this base pairing.

The nitrogenous bases form hydrogen bonds with one another, holding the two DNA strands together. Because the base pairing is distinct and complementary, it is possible to predict the sequence of one strand from that of the other. The transcription and replication of DNA depend on this complementary base pairing.

Cells duplicate their DNA before dividing, a process known as DNA replication. The two DNA strands split apart during replication, and each strand acts as a model for the creation of a new complementary strand. DNA polymerases, which add nucleotides to the new strand in accordance with the laws of base pairing, catalyze this process.

Messenger RNA (mRNA), which conveys the genetic information from the DNA to the ribosomes where proteins are made, is created using DNA as a template during transcription. An enzyme known as RNA polymerase copies a segment of DNA into mRNA during transcription. The mRNA and DNA template are complementary, but the mRNA substitutes uracil (U) for thymine (T).

The order of nucleotides along the DNA strands encodes the genetic information. The sequence of amino acids in proteins is defined using this information. The building blocks of proteins are amino acids, and the order of the nucleotides in the DNA determines the specific amino acid sequence of each protein.

The set of guidelines known as the genetic code describes how the sequence of amino acids in proteins and nucleotides in DNA relate to one another. Codons, which are triplets of nucleotides that the ribosomes read while constructing proteins, are the foundation of the genetic code. Since there are only 20 amino acids and 64 potential codons, most amino acids are designated by more than one codon.

Gene expression, or the process by which genes are switched on or off in response to environmental or developmental cues, is regulated in part by DNA. A complex network of regulatory elements, such as promoters, enhancers, and silencers, interacts with proteins to regulate the transcription of certain genes, regulating gene expression.

DNA may change on its own or as a consequence of exposure to mutagens like radiation or certain substances. Proteins' amino acid sequences may change as a result of mutations, changing both the structure and the function of the protein. While certain mutations may be dangerous or fatal, others may be advantageous.

In conclusion, DNA's structure and function are crucial for the growth, development, and functionality of all living things. Complementary base pairing, which is necessary for DNA replication and transcription, is the basis of the double helix structure of DNA. The genetic code provides the connection between the sequence of nucleotides in DNA and the sequence of amino acids in proteins, and DNA encodes genetic information that is utilized to determine the sequence of amino acids in proteins. Additionally, DNA is essential for controlling gene expression, and DNA mutations

The Discovery of DNA

One of the most significant scientific discoveries in biology's history is the discovery of DNA. It marks a significant turning point in the study of life itself, the growth of biotechnology, and our knowledge of genetics. A lot of scientists spent several years researching this finding, expanding on the work of their forerunners.

When researchers started looking into the characteristics of cells and the components they contained in them in the late 19th century, they made the discovery of DNA. Early in the 20th century, scientists started to pay attention to nucleic acids, the intricate chemicals contained in cells' nuclei. They were aware that the four distinct chemical bases that made up these molecules were there, but they had no idea of the composition, structure, or purpose of these molecules.

Frederick Griffith, a British scientist, performed a ground-breaking experiment on bacteria in 1928 that illustrated the idea of transformation—the idea that a material in one organism may modify the characteristics of another. Griffith was researching two types of pneumonia-causing bacteria, one of which was virulent and the other not. Mice perished when he injected them with the virulent strain, while the non-virulent strain did no damage. However, the mice died, and he discovered surviving virulent bacteria in their tissues when he injected them with a combination of heat-killed virulent bacteria and living non-virulent bacteria. This indicated that the non-virulent bacteria had acquired some property from the deceased virulent bacteria, rendering them virulent.

In spite of the fact that it was still unclear what this genetic substance was, this experiment offered the first concrete proof that it existed. Griffith's discoveries were utilized as a springboard by a group of scientists headed by Oswald Avery, Colin MacLeod, and Maclyn McCarty in the 1940s for their own investigation. They discovered that only DNA, not RNA or proteins, was capable of converting bacteria using a technique identical to Griffith's but using pure biological components. This gave compelling proof that DNA made up the genetic material.

Rosalind Franklin, a graduate student at King's College in London, started studying the structure of DNA in 1952 using a method called X-ray crystallography, which included exposing a crystal of the molecule to X-rays and examining the pattern of diffracted rays that resulted. The molecule seemed helical in Franklin's photographs, indicating that it had a regular repeating structure.

Two young scientists from the University of Cambridge who were also interested in the structure of DNA at the time were James Watson and Francis Crick. To create a model of the molecule, they combined data from other scientists with information from Franklin's photos. They suggested that the nitrogenous bases were coupled in certain combinations (adenine with thymine and guanine with cytosine, for example) that allowed for complementary base pairing and that the DNA structure was a double helix with two strands of nucleotides coiled around one another.

One of the most significant scientific articles in history was the one that Watson and Crick published in the magazine Nature in 1953. Their knowledge of the structure of DNA helped frame studies in genetics, biochemistry, and molecular biology and opened up new directions for investigation.

In the years that followed, researchers furthered their comprehension of DNA and developed new approaches for examining its structure and function. This sparked the establishment of brand-new scientific disciplines like genetic engineering and genomics, which have fundamentally changed how we see genetics and the function of DNA in all living things.

In conclusion, a new era in genetics study began with the discovery of DNA, which was a significant turning point in the history of biology. Genetics, biochemistry, and molecular biology have all significantly advanced as a result of the discovery of DNA, which offered the first conclusive proof of the presence of a hereditary substance.

We continue to investigate the multiple functions that DNA plays in the growth, development, and operation of all living things since we now have a far better knowledge of the structure and function of DNA. The discovery of DNA is a monument to the force of scientific inquiry and the quest for knowledge, ranging from the early work of Griffith and Avery to the ground-breaking studies of Franklin, Watson, and Crick. It has had a significant influence on how we see the natural world and has opened up new vistas in our search to understand the origins of life.

The Structure of DNA

All living things include the lengthy, double-stranded molecule known as deoxyribonucleic acid (DNA), which also happens to have a helical structure. James Watson and Francis Crick initially identified DNA's structure in 1953, and their discovery of the double helix structure had a profound impact on how we understood genetics and molecular biology.

A nitrogenous base, a sugar with five carbons termed deoxyribose, and a phosphate group make up the fundamental components of DNA, known as nucleotides. DNA contains adenine (A), thymine (T), guanine (G), and cytosine (C), four distinct forms of nitrogenous bases. In the pairing of the nitrogenous bases, A and G always pair with T and C, respectively, to make up the fundamental components of DNA, known as nucleotides. DNA contains adenine (A), thymine (T), guanine (G), and cytosine (C), four distinct forms of nitrogenous bases. In the pairing of the nitrogenous bases, A and G always pair with T and C, respectively. Hydrogen bonds that develop between the nitrogenous bases hold these base pairs together.

The 5' end of one strand is close to the 3' end of the other strand since the two strands of the DNA double helix run in opposing orientations. The two strands' sugar-phosphate backbones also run in opposing directions, with one strand's 5' end being close to the other's 3' end.

Hydrogen bonds between the base pairs keep the double helix's two strands together. Adenine and thymine make a single hydrogen bond, but guanine and cytosine form a stronger double bond. This implies that the amounts of A and G are always identical to one another, as well as the amounts of T and C, respectively. Chargaff's rule is what we refer to it as.

Numerous interactions between the various parts of the molecule help to stabilize the double helix of DNA. The two strands of the helix are held together by hydrogen bonds between the base pairs, and further stability is provided by interactions between the sugar and phosphate groups.

In addition to its fundamental structure, DNA has a variety of additional characteristics that are crucial for its function. For instance, telomeres, which shield the DNA from deterioration and damage, cap the ends of each strand of DNA. Additionally, DNA is contained in a sophisticated structure known as chromatin, which aids in organizing the DNA and managing its accessibility to the cellular machinery responsible for reading and copying it.

In order to perform its crucial function in the storage and transfer of genetic information, DNA's structure has developed over millions of years into a very sophisticated and complicated molecular architecture. Modern genetics and molecular biology have advanced significantly as a result of the understanding of DNA structure, which was a crucial milestone in this progress.

DNA Replication and Transcription

Transcription and DNA replication are two essential processes for the transfer of genetic information from one cell generation to the next. These strictly regulated processes need the coordinated action of several unique enzymes and proteins.

A cell duplicates its DNA via a process known as DNA replication prior to cell division. During this procedure, the two strands of the DNA double helix are separated by an enzyme called helicase, which dissociates the hydrogen bonds between the base pairs and unwinds the DNA molecule. After that, the enzyme DNA polymerase utilizes the split strands as templates to start over with new complementary strands.

DNA polymerase adds nucleotides to the growing strand in a specific sequence while adhering to the complementary base pairing rules. The new strand is synthesized in the 5' to 3' direction, whereas the template strand is read in the 3' to 5' direction. This means that DNA polymerase can only bind nucleotides to the growing strand in the 5' to 3' direction.

The starting and stopping points of replication are known as the origins of replication. The replication initiation complex, a group of proteins, is in charge of identifying these replication origins. Once replication has started, DNA polymerase moves along the template strand while continually building new strands on the leading strand and sporadic new strands on the trailing strand. The lagging strand's irregularly formed parts are known as Okazaki fragments, and DNA ligase is what eventually connects them.

DNA Transcription: Transcription is the process of utilizing the genetic code of DNA to produce messenger RNA (mRNA), which carries information from the nucleus to the cytoplasm, where it is translated into proteins. An enzyme known as RNA polymerase, which is involved in transcription, binds to the promoter, a specific region of DNA, and begins to unwind the double helix.

After reading the template strand of DNA in the 3' to 5' direction, RNA polymerase synthesizes the equivalent RNA molecule in the 5' to 3' direction. As it moves along the DNA template, RNA polymerase produces an RNA molecule that is complementary to the DNA template strand.

The RNA molecule produced during transcription is processed in a variety of ways before it is removed from the nucleus. For instance, the splicing process removes introns, which are non-coding RNA segments. Joining the remaining exons results in a mature mRNA molecule.

After mature mRNA molecules leave the nucleus and enter the cytoplasm, ribosomes convert them into proteins. The nucleotide sequence of the mRNA molecule determines the amino acid composition of the protein.

Overall, DNA replication and transcription are two key processes that are necessary for the transfer of genetic information from one generation of cells to the next. These strictly regulated processes need the coordinated action of several unique enzymes and proteins. Our comprehension of these mechanisms has considerably benefited modern genetics and molecular biology and resulted in several advancements in these fields.

Chapter 4:

Genetic Mutations and Disease

A genetic mutation is a modification to the DNA sequence of an organism. These alterations might have a range of effects, from nothing at all to the emergence of serious diseases. A genetic mutation may be passed down to a child from one or both parents, or it might appear spontaneously during cell division or as a result of exposure to environmental triggers.

Genetic alterations may disrupt the way that genes function properly, which may result in diseases. Mutations may affect both the manufacturing and operation of proteins because they can alter the DNA sequence's instructions. This might alter cellular functions that are crucial to usual biological processes, including cell division, differentiation, and death.

The two kinds of mutations that may occur are somatic and germline. Germline mutations occur in reproductive cells, while somatic mutations occur in the body's non-reproductive cells and are not passed on to the next generation. Diseases like cancer may result from somatic mutations.

There are many distinct types of genetic mutations, including point mutations, insertions, deletions, duplications, and translocations. Point mutations are alterations in a single nucleotide, as opposed to insertions and deletions, which involve the addition or deletion of nucleotides. While duplications require the replication of a piece of DNA, translocations involve the movement of a piece of DNA from one position to another.

Among the diseases that may arise as a consequence of faulty genes are genetic disorders including cystic fibrosis, sickle cell anemia, and Huntington's disease. Furthermore, they may contribute to the development of major conditions including cancer, heart disease, and Alzheimer's. The kind and location of the mutation influence the severity of the illness.

To create treatments for genetic diseases, it is critical to understand genetic changes and how they impact common biological activities. Advances in genetic research and technology have led to the development of personalized medicine, which involves tailoring medications to a person's specific genetic profile. This could improve the effectiveness and lessen the side effects of treatments for inherited diseases.

Types of Genetic Mutations

Changes in the DNA sequence known as genetic mutations may happen either spontaneously or as a result of exposure to certain environmental conditions. Genetic mutations may take many different forms, each with its own properties and impacts on the organism. Understanding how these various kinds of mutations affect human health and the emergence of illnesses is essential.

Changes at a Point

A single nucleotide in the DNA sequence may change in a point mutation. Point mutations come in three different flavors: substitutions, insertions, and deletions. While insertions and deletions require the addition or removal of one or more nucleotides, substitutions entail the replacement of one nucleotide with another. Point mutations may have little effect on the organism or may result in significant genetic disorders, among other outcomes.

Mutations in the chromosome

Changes to a cell's chromosome structure or number are known as chromosomal mutations. These mutations may have a substantial effect on the growth and survival of the organism.

Chromosome mutations come in a variety of forms, including deletions, duplications, inversions, and translocations. While duplications entail the replication of a portion of a chromosome, deletions involve the loss of a portion of the chromosome. Translocations entail the migration of a segment of a chromosome from one place to another, while inversions require the reversal of a section of a chromosome.

Mutations in Trinucleotide Repeat Expansion

Trinucleotide-repeat expansion mutations happen when a three-nucleotide sequence repeats itself repeatedly inside a gene. These mutations may have no impact at all or result in significant genetic disorders like Huntington's disease and fragile X syndrome, among other outcomes for the organism.

Mutations resulting from gene duplication

When a piece of DNA is copied and introduced into the genome, creating an additional copy of the gene, gene duplication mutations take place. Increased protein synthesis as a result of these mutations may have a variety of repercussions on the organism.

Changes in Frames

Frameshift mutations take place when nucleotides are either added to or taken out of the DNA sequence, shifting the reading frame. This may result in modifications to the protein's amino acid sequence, which may materially alter how well the protein performs.

Mutations at the Splice Site

Splice site mutations, which impact the splicing of the mRNA and the creation of the protein, occur at the junction between introns and exons in the DNA sequence. Genetic disorders like beta-thalassemia and cystic fibrosis may develop as a result of these mutations.

For the diagnosis and treatment of genetic illnesses, it is crucial to comprehend the many kinds of genetic mutations and how they affect an organism. The creation of personalized medicine, which includes adapting therapies to a person's unique genetic profile, is a result of advances in genetic research and technology. This may increase the efficiency and decrease the negative effects of therapies for hereditary illnesses.

Genetic Diseases and Disorders

Genetic mutations or DNA sequence changes that result in faulty proteins or other cellular functions are the basis for genetic illnesses and disorders. The DNA of a person may naturally include these mutations or acquire them from their parents.

There are hundreds of recognized hereditary illnesses and disorders, each with its own features and physiological implications. A single gene mutation may cause certain genetic illnesses, but several gene mutations or chromosomal abnormalities can cause other genetic disorders. The following are some of the most prevalent hereditary illnesses and disorders:

Dysplastic fibrosis

Mutations in the CFTR gene, which produces a protein that controls the movement of salt and water into and out of cells, result in cystic fibrosis, a genetic condition. The mutations produce thick, gummy mucus that clogs the lungs and other organs and causes digestive issues, respiratory infections, and other consequences.

Acute myeloid leukemia

Mutations in the HBB gene, which produces the beta-globin protein that is a component of hemoglobin, lead to sickle cell disease, a hereditary illness. The abnormalities result in sickle- or crescent-shaped red blood cells, which may obstruct blood vessels and decrease the quantity of oxygen that reaches body tissues, resulting in anemia, discomfort, and other consequences.

Hemophilia

Mutations in the genes that produce clotting factors, which are proteins that aid in blood clotting, result in hemophilia, a hereditary condition. In hemophiliacs, blood clots cannot properly form, which may result in excessive bleeding, bruising, and, in extreme instances, life-threatening hemorrhage.

Dwarf Syndrome

An extra copy of chromosome 21 causes the genetic condition known as Down syndrome. Intellectual impairment, developmental delays, and distinctive physical traits including poor muscle tone, almond-shaped eyes, and a single crease across the palm of the hand are the effects of this.

Alzheimer's disease
A mutation in the HTT gene, which produces the huntingtin protein, results in Huntington's disease, a hereditary illness. The mutation causes mobility impairments, cognitive decline, and psychiatric symptoms by causing the nerve cells in the brain to degenerate.

Skeletal dystrophy
A series of hereditary illnesses known as muscular dystrophy result in gradual muscle deterioration and weakening. Mutations in the genes that create the proteins required for muscle function, such as dystrophin in Duchenne muscular dystrophy, are the root cause of these diseases.

PKU, or phenylketonuria,
Mutations in the PAH gene, which codes for the enzyme required to break down the amino acid phenylalanine, result in phenylketonuria, a hereditary condition. Without this enzyme, the body may accumulate phenylalanine and its metabolites, which can cause intellectual impairment, developmental delays, and other issues.

Tay-Sachs condition

Mutations in the HEXA gene, which codes for an enzyme that breaks down a form of fat known as gangliosides, result in Tay-Sachs disease, a hereditary condition. Without this enzyme, gangliosides may build up in nerve cells and cause seizures, visual and hearing loss, developmental delays, and other issues.

Turner disease

When the X chromosome is absent entirely or partially in females, Turner syndrome develops. Short height, problems with reproduction, and other physical and developmental difficulties might result from this.

Williams disease

A loss of genetic material on chromosome 7 results in Williams syndrome, a genetic condition. Developmental delays, intellectual incapacity, and distinctive physical traits, including an "elfin" face look and cardiovascular issues, may come from this.

Genetic testing, which may spot particular mutations or chromosomal abnormalities, is often used in the diagnosis of genetic illnesses and disorders. Depending on the particular illness and its severity, different genetic diseases and disorders have different treatment choices. Some genetic diseases are incurable, so therapy focuses on symptom management and avoiding consequences. However, developments in genetic therapy and research have resulted in the creation of fresh remedies for several hereditary illnesses, such as:

Replacement of enzymes

Some hereditary illnesses caused by a lack or shortage of a particular enzyme are treated using enzyme replacement therapy. In order to replace the defective enzyme and enhance the body's capacity to break down certain chemicals, this treatment entails injecting the patient's bloodstream directly with the missing enzyme.

Gene treatment

In order to replace or repair a damaged gene, gene therapy involves making changes to the DNA of a patient's cells. Viral vectors, which are altered viruses that can carry a functioning gene to the patient's cells, may be used to do this. Although gene therapy is still in its early stages, it has shown promise in treating several hereditary illnesses.

Transplanting stem cells

Replacement of the patient's bone marrow or blood-forming cells with healthy donor cells is the goal of stem cell transplantation. This is applicable to the treatment of thalassemia and sickle cell disease, two hereditary conditions that damage the blood cells.

Pharmaceutical treatments

Pharmacological treatments utilize medications to treat symptoms and stop the progression of genetic illnesses. For instance, those who have cystic fibrosis could take drugs to decrease mucus, enhance lung function, and guard against infections.

Diet modifications

Changes in nutrition may help treat certain hereditary illnesses, such as PKU. A low-phenylalanine diet may stop the body from accumulating phenylalanine in the case of PKU and stop developmental delays.

In addition to these therapies, genetic counseling is a crucial component in managing genetic illnesses and disorders. Working with a qualified practitioner to explore testing alternatives, address the emotional and practical ramifications of a genetic diagnosis, and explain the dangers of passing on a genetic illness to kids is known as genetic counseling.

The overall complexity of genetic illnesses and disorders may have a major influence on a person's health and wellbeing. New therapies and better results for people suffering from these disorders are possible because of advancements in genetic research and therapy.

Gene Therapy and the Future of Disease Treatmen

The use of functioning genes to replace or fix damaged genes inside a person's DNA is a promising strategy for treating genetic diseases. It is a method for treating, curing, or preventing illnesses by changing the genes that lead to certain ailments. Gene therapy has a bright future in the treatment of many illnesses that were previously considered to have no cure. This technology has the potential to completely alter healthcare and how we treat diseases.

Gene therapy's main objective is to repair genetic problems by inserting functioning genes into a patient's DNA to replace or fix damaged genes. Somatic gene therapy and germ-line gene therapy are two of the several methods used in gene therapy. Genetic diseases in non-reproductive cells, such as blood or muscle cells, are treated via somatic gene therapy. On the other side, germ-line gene therapy modifies genes in reproductive cells like sperm or egg cells, which may pass on the improved genes to subsequent generations.

Gene therapy comes in a variety of forms, such as gene substitution, gene editing, and gene suppression. Gene replacement treatment involves the introduction of functioning genes to take the place of damaged genes. To fix a genetic problem, gene editing entails changing the DNA sequence inside a cell. The process of suppressing the expression of a problematic gene is known as gene suppression treatment, often referred to as RNA interference.

One of gene therapy's most important benefits is its ability to cure hereditary illnesses that are either incurable or have few other effective therapeutic alternatives. For instance, the digestive and respiratory systems are both impacted by the hereditary condition cystic fibrosis. There is no known therapy for cystic fibrosis, and there are few available treatments. Correcting the genetic flaw that leads to cystic fibrosis using gene therapy might lead to the disease's eventual recovery.

Gene therapy also has the ability to treat certain illnesses in a tailored manner, which is a huge benefit. Traditional pharmacological interventions often have systemic negative effects and may be quite harmful. The efficiency of treatment may be increased and the danger of adverse effects decreased using gene therapy, which can be tailored to target certain cells or organs.

Gene therapy has possibilities, but it also has drawbacks. The transport of genes to the target cells or tissues is one of the major obstacles. There are several ways to distribute genes, including physical techniques, non-viral vectors, and viral vectors. The most popular technique for delivering genes is using viral vectors, but these might result in immunological reactions and other issues. On the other side, non-viral vectors are less effective yet safer.

The potential for off-target effects, in which the injected genes may mistakenly alter other genes in the genome, is another difficulty with gene therapy. Unintended effects might result in the emergence of new illnesses or the escalation of current ones. To reduce the possibility of off-target consequences, scientists are attempting to create safer and more accurate gene editing techniques.

Despite these difficulties, gene therapy has enormous potential for the treatment of many different illnesses. Gene therapy has advanced significantly in recent years, with the US Food and Drug Administration (FDA) approving the first gene therapy product in 2017. Acute lymphoblastic leukemia is a type of blood cancer, and the product Kymriah is a gene therapy treatment for it.

In conclusion, gene therapy has enormous promise for treating a wide range of illnesses that were previously deemed incurable. While there are certain difficulties with gene therapy, such as the potential for off-target effects and the transport of genes to target cells, developments in the field are making the procedure safer and more efficient. The potential for gene therapy to transform healthcare and alter how we treat diseases makes the field's future very promising.

Chapter 5:

Genomics and Personalized Medicine

Medicine

The study of an organism's genetic makeup, or DNA, is known as genomics. It entails the sequencing and examination of a person's whole genome, which is where all of their genetic information is located.

Contrarily, personalized medicine is a kind of medical care that offers customized treatment choices by taking into consideration a patient's particular genetic make-up, lifestyle, and environment. Through the development of more precise and efficient therapies for a variety of illnesses, the marriage of genomics and personalized medicine has the potential to completely transform healthcare.

Healthcare professionals may learn more about a patient's propensity for certain illnesses by looking at their genetic data and then modify treatment methods accordingly. This strategy may result in therapies that are more successful and have fewer negative effects. The merging of genomics and personalized medicine is becoming more accessible and has the potential to greatly enhance patient outcomes as the cost of genome sequencing continues to fall.

The Human Genome Project

Aiming to discover and sequence every gene in the human genome, which is the whole collection of genetic instructions that make up an individual, the Human Genome Initiative (HGI) was an international scientific research initiative. The Human Genome Project (HGP) began in 1990 and ended in 2003, with the announcement of the entire human genome sequence occurring in April 2003. It took hundreds of scientists from across the globe working together on the project.

Each human cell has 23 pairs of chromosomes, which together make up the approximately 3 billion base pairs that make up the human genome. The objective of the HGP was to map and sequence all of these base pairs and determine the 20,000–25,000 genes that make up the human genome. To do this, the team used a number of technologies, including computer analysis, automated sequencing equipment, and DNA fragment mapping.

The Human Genome Project (HGP) set out to study the function of genes and how they relate to illness. Among its many objectives were to identify every gene in the human genome and determine its sequence. The creation of novel technology and techniques for DNA sequencing, which have transformed genetics research and made genome sequencing more available and inexpensive, was one of the HGP's most important accomplishments.

The HGP has significantly influenced both biology and medicine. The research yielded a plethora of knowledge on the composition and operation of the human genome, including the discovery of genes that cause illness and genetic variants that influence a person's propensity to get certain illnesses. This knowledge has sparked the creation of novel diagnostic procedures and therapeutic approaches for a variety of illnesses, including cancer, heart disease, and genetic abnormalities.

In addition to its medicinal uses, the HGP has revealed fresh perspectives on the development of people and the history of our species. The experiment showed that humans, chimpanzees, and other primates all had a common progenitor and that our species started in Africa some 200,000 years ago.

There were issues with the HGP. Critics said that the study was costly and that the funds would have been better used for other medical studies or public health programs. Others expressed moral concerns about the possible abuse of genetic data and the privacy and discrimination implications of genetic testing.

The HGP continues to be among the most important scientific breakthroughs of the contemporary period despite these worries. The endeavor opened the door for fresh insights and advancements in genomics study and customized therapy. The legacy of the HGP has created new opportunities for the detection and treatment of illness and continues to influence how we understand genetics and human biology.

Genome Sequencing and Analysis

The process of finding and analyzing the genetic data included in an organism's DNA is known as genome sequencing and analysis. Genome sequencing is the process of deciphering the nucleotide base order in an organism's DNA, which reveals information about the genetic codes responsible for a person's physical and biological characteristics. Genome analysis is evaluating this data in order to comprehend how genes work and how they affect both health and sickness.

DNA extraction, fragmentation, amplification, and sequencing are all phases in the sequencing of a genome. DNA extraction entails separating the DNA from the organism's cells for sequencing. The process of fragmentation includes cutting the DNA into smaller fragments, each of which is subsequently duplicated by amplification. The amplified fragments are then sequenced using cutting-edge methods like PacBio sequencing or Illumina sequencing. A whole genome sequence is subsequently created by assembling the resultant sequences.

Genomic analysis entails analyzing the genomic sequence data using computer techniques. Genes and other functional components of the genome may be identified; the genome's sequence can be compared to those of other genomes to determine its similarities and differences; and the genome can be examined for mutations that could be linked to illness or other features.

In the area of personalized medicine, genome sequencing and analysis have several important applications. Healthcare professionals may learn more about a patient's propensity for certain illnesses by studying their genome and can then modify treatment regimens appropriately. For instance, if a genetic test reveals that a person is more likely to get breast cancer, this may lead to earlier and more frequent cancer screenings.

Understanding the genetic underpinnings of illnesses and other biological events also uses genome sequencing and analysis. For instance, to find potential genetic variants linked to an illness, researchers may compare the genomes of people with and without a certain ailment. The science of synthetic biology also uses genome analysis to create organisms with desirable characteristics, such as better disease resistance or higher production.

Recent advancements in sequencing technology and the creation of computer tools for processing genome data have made genome sequencing and analysis more widely available. There are still a lot of obstacles to be overcome, such as the need for more precise and dependable sequencing technology, better resources for processing and deciphering genome data, and more efficient ways to handle and store massive volumes of genomic data.

Overall, genome sequencing and analysis have the promise of revolutionizing both medical practice and biological knowledge. There will probably be new discoveries and advancements as the subject develops that will help society in a variety of ways.

The Potential for Personalized Medicine.

The discipline of personalized medicine is expanding quickly and aims to customize medical care to a person's particular genetic make-up, environment, and lifestyle. Personalized medicine aims to lessen possible negative effects while increasing the efficacy of medical therapies. This method is distinct from conventional medicine, which uses the same treatment plan for all patients.

Personalized medicine has enormous promise. Doctors can more accurately forecast which therapies will be most beneficial and which may have unfavorable side effects by analyzing a patient's genetic composition. This information may lessen the need for trial and error throughout the course of therapy by assisting clinicians in prescribing the appropriate medication at the appropriate dosage.

Personalized medicine has emerged as a result of advances in genomics. In the past, medical professionals treated patients uniformly. To choose the most effective course of action for a particular illness, they looked to empirical data and clinical studies. However, recent developments in genomics technology have made it feasible to tailor medical care based on a patient's genetic profile.

Researchers have discovered specific genes linked to certain medical disorders via the study of genomics. They have also discovered genetic differences that may affect a person's reaction to a particular medicine. With this information, medical professionals may create custom treatment strategies that consider each patient's particular genetic profile.

The ability to lessen the possibility of negative medication responses is one of the main advantages of customized medicine. Traditional medicine has a serious issue with adverse medication responses. It may be challenging for physicians to identify which pharmaceuticals will result in unpleasant side effects for their patients. With customized medicine, physicians may modify a patient's treatment regimen depending on their genetic profile, lowering the risk of negative side effects.

The treatment of cancer using personalized medication has already shown potential. Researchers have discovered certain genetic alterations that are connected to various cancer types. Doctors may create tailored medicines that attack cancer cells while sparing healthy cells by learning about a patient's genetic make-up. Some cancers, including lung and breast cancer, have responded well to this kind of treatment.

Treatment of infectious illnesses is a further potential area for customized medicine. Doctors may be able to create tailored medicines that go after the particular strain of bacteria or virus causing the sickness by researching a patient's genetic composition. This strategy could lessen the need for broad-spectrum antibiotics, which can lead to the emergence of bacteria that are resistant to antibiotics.

Personalized medicine may provide advantages, but there are also issues that need to be resolved. The price of genetic testing is one of the main obstacles. Genetic testing is currently costly, and not all patients' insurance plans will always cover the expense. Patients may find it challenging to get tailored treatment as a result.

The interpretation of genetic data presents another difficulty. Since genomics is still a young discipline, there are still many aspects of the human genome that are unknown to scientists.

As a consequence, deciphering genetic information may be difficult and complicated. For the purpose of creating efficient, individualized treatment programs, doctors must possess the knowledge and training to appropriately analyze genetic data.

In conclusion, individualized medicine has enormous promise. Doctors may create individualized treatment strategies that are more successful and have fewer negative effects by studying a patient's genetic composition. The advantages of customized medicine are substantial and have the potential to completely transform the practice of medicine, even if there are certain obstacles that must be overcome. We may anticipate future advancements in customized medicine that will enhance patient outcomes and lessen the burden of illness as the field of genetics develops.

Chapter 6:

Genetic Engineering and Biotechnology

Two fast-developing scientific areas, genetic engineering and biotechnology, have the power to drastically alter many facets of our existence. These domains are fundamentally concerned with the modification of genetic material to produce new or altered molecules, organisms, or metabolic pathways.

By adding new DNA or changing existing DNA, genetic engineering is the act of changing an organism's genetic make-up. Using this method, researchers may add or delete certain qualities from an organism's genetic composition to produce new varieties with the necessary properties. The development of novel medical procedures, the improvement of food items, and the development of pest- and disease-resistant crops have all been made possible via genetic engineering.

On the other hand, biotechnology uses live things like cells, enzymes, or living beings to create commodities and services. Applications in this area span widely, from the creation of biofuels to the creation of fresh vaccinations and treatments. The use of genetic engineering methods to develop new and enhanced biological products is included in the field of biotechnology.

Both biotechnology and genetic engineering have the potential to completely transform many facets of our existence. They provide fresh answers to some of the most urgent issues facing the globe today, such as hunger, sickness, and environmental destruction. The production of genetically modified organisms (GMOs) and the possible hazards connected to their release into the environment, however, present additional ethical and security issues in these domains.

In general, genetic engineering and biotechnology are fascinating, rapidly developing subjects with a bright future. These sectors are expected to play a bigger part in forming our reality and solving some of the most serious problems we face as long as research and development are done in them.

Genetic Modification of Crops and Animals

Genetic modification (GM) of crops and animals is a biotechnology technique that involves the deliberate modification of an organism's genetic material to create desired traits. GM crops and animals are developed using advanced molecular biology techniques, including genetic engineering and genome editing. These techniques allow scientists to introduce specific genes or modify existing genes to produce organisms with desirable characteristics, such as increased yield, pest resistance, or improved nutritional content.

GM crops are widely used in agriculture with the aim of increasing crop yields, reducing the use of pesticides, and improving the nutritional quality of food. The most commonly genetically modified crops include corn, soybeans, cotton, and canola. GM crops are created by inserting a gene from another organism, such as a bacteria or virus, into the plant's DNA. This gene may provide the plant with a new trait, such as resistance to a particular herbicide or the ability to produce a natural pesticide.

GM animals are also being developed for a variety of purposes, including improving livestock productivity, developing new medical treatments, and studying disease. For example, GM salmon have been created to grow faster and require less feed, potentially reducing the environmental impact of salmon farming. GM mice have been created for use in medical research, providing a valuable tool for studying human diseases and developing new treatments.

While genetic modification of crops and animals has many potential benefits, there are also concerns about the potential risks associated with this technology. One major concern is the possibility of unintended consequences, such as the spread of GM traits to wild relatives or the creation of new diseases. There are also concerns about the impact of GM crops on the environment and human health.

To address these concerns, many countries have established regulatory frameworks to oversee the development and use of GM crops and animals. These frameworks require that GM crops and animals undergo rigorous testing to ensure that they are safe for human health and the environment. In addition, many companies that develop GM crops and animals have established guidelines for the ethical and responsible use of this technology.

Overall, genetic modification of crops and animals is a rapidly evolving field that has the potential to provide many benefits to society. However, it is important that this technology be used responsibly and with careful consideration of the potential risks and benefits. By continuing to study and develop GM crops and animals in a responsible manner, we can potentially address many of the world's most pressing challenges, from food security to disease prevention.

The Ethics of Genetic Engineering

One area of science that is quickly developing and has the potential to significantly alter our lives is genetic engineering. However, there are a number of ethical issues with this technology's usage and applications. The ethics of genetic engineering are intricate and varied, taking into account social advantages, individual rights, and possible hazards.

The possibility of unforeseen repercussions is one of the main ethical issues with genetic engineering. Genetically modified organisms (GMOs) run the danger of having adverse effects on the environment or on people's health since the results of the genetic alteration of organisms might be unexpected. For instance, adding new features to crops or animals could have unforeseen effects on the interactions between those species and other species. GMOs may also contribute to the emergence of new illnesses or antibiotic resistance, which is another worry.

The potential for "designer babies" or genetically altered people is another ethical concern associated with genetic engineering. Utilizing this technology, one might choose kids based on attributes like athletic prowess, intellect, or physical attractiveness. This raises questions about the likelihood of prejudice and inequity as well as the danger it might pose to the people involved.

Concerns exist over how genetic engineering may affect natural ecosystems and biodiversity. The loss of genetic variety brought on by the genetic alteration of organisms may reduce the ability of natural systems to withstand environmental shocks. Furthermore, there is concern that genetic engineering can result in the development of invasive species that harm ecosystems.

Many nations have created regulatory frameworks for the use of genetic engineering in order to address these ethical issues. Before every novel genetic change is authorized for use, these frameworks seek to guarantee that genetic engineering is handled responsibly and that any dangers are thoroughly assessed. A lot of businesses that create genetically modified goods have also set rules for the moral and responsible use of this technology.

In general, the ethics of genetic engineering are intricate and varied, taking into account social advantages, individual rights, and possible hazards. It is crucial that we keep having meaningful ethical debates regarding the usage and applications of genetic engineering and that we develop this technology with caution and responsibility. By doing this, we may be able to maximize the advantages of genetic engineering while limiting its dangers and detrimental effects.

Although the area is complicated and there are several more ethical difficulties that occur in the context of genetic engineering, the following summary gives a comprehensive overview of the main ethical concerns associated with genetic engineering.

For instance, the potential for genetic engineering to aggravate societal inequities is another ethical problem. Genetic engineering may cause the income gap between rich and poor to increase if technology is only accessible to the affluent. There are also worries about the possibility of using genetic engineering in ways that are harmful to people or society at large, such as the creation of bioweapons.

Concerns exist over the application of informed consent to genetic engineering. Genetically modified people could not completely comprehend the possible hazards and advantages of the operation, and there is a possibility that they might be forced to undergo the treatment without giving their full, informed consent.

Last but not least, there are worries about the possibility of culturally incorrect or disrespectful uses of genetic engineering. For instance, it can be considered disrespectful or even insulting to employ genetic engineering to alter crops or animals that are holy or important to some people on a cultural level.

In conclusion, there are several ethical concerns that emerge in the context of genetic engineering and that are complicated and multidimensional. It is crucial that we keep having meaningful ethical debates regarding the usage and applications of genetic engineering and that we develop this technology with caution and responsibility. By doing this, we may be able to maximize the advantages of genetic engineering while limiting its dangers and detrimental effects.

CRISPR and the Future of Genetic Modification

The powerful new genetic editing tool CRISPR (Clustered Regularly Interspaced Short Palindromic Repeats) has the potential to transform a wide range of scientific and medical fields. The CRISPR system has been compared to a pair of molecular scissors that can be used to cut and paste DNA with unmatched accuracy, enabling scientists to precisely alter the genes of living creatures.

The first instance of CRISPR was found in bacteria, where it functions as a viral defense. In order to target and eliminate viruses they come across again, bacteria utilize CRISPR to store genetic information about viruses they have already seen. Researchers discovered that they could use CRISPR to modify genes in other creatures, creating new opportunities for genetic engineering.

The realm of medicine has one of the most interesting uses for CRISPR technology. By altering the defective genes that cause hereditary disorders like sickle cell anemia and cystic fibrosis, scientists are investigating the use of CRISPR to cure these ailments. In experiments on animals, CRISPR has been effective in treating conditions including Duchenne muscular dystrophy and Huntington's disease. But before CRISPR can be utilized in a therapeutic context, there are a number of technological and ethical issues that must be resolved. The use of CRISPR in human patients is still in the early phases of research.

Agriculture is a sector where CRISPR might also find use. With CRISPR, it may be possible to develop crops with better yields or greater disease resistance, thus boosting food security and lowering the need for pesticides. Genetically modified crops might have a negative influence on the environment, so before using this technology widely, it is vital to carefully weigh the risks and advantages.

Important ethical issues about the proper use of genetic alteration are also raised by the use of CRISPR. The broad use of CRISPR has raised concerns that it can have unforeseen repercussions, such as the emergence of new illnesses or the production of genetically altered creatures that might have a severe influence on the environment. Furthermore, there are worries that CRISPR might be used in ways that are harmful to people or society at large, such as the development of bioweapons or the production of "designer babies."

Many nations have created regulatory frameworks for the use of CRISPR and other genetic alteration technologies in order to address these ethical issues. Before every novel genetic change is authorized for use, these frameworks seek to guarantee that genetic engineering is handled responsibly and that any dangers are thoroughly assessed.

In conclusion, CRISPR is a formidable new tool for genetic alteration that might completely alter several fields of research and medicine. However, the introduction of CRISPR also prompts significant moral issues about the appropriate use of genetic manipulation. It is crucial that we continue to have in-depth ethical conversations regarding the usage and applications of CRISPR and that we develop this technology cautiously and responsibly. By doing this, we may be able to maximize the advantages of CRISPR while reducing its dangers and detrimental effects.

The information above gives a comprehensive introduction to CRISPR technology, including its potential uses in agriculture and medicine as well as the ethical issues and legal frameworks that surround its usage. However, the field of genetic modification is quickly developing, and there are continuous research and development projects to enhance CRISPR technology and investigate its possible uses in other fields.

For instance, researchers are looking at how CRISPR may be used to produce novel cancer treatments that would specifically target and eliminate cancer cells while leaving healthy ones unharmed. Additionally, there are initiatives to employ CRISPR to produce new substances and molecules with unique features, such as biodegradable polymers or fresh antibiotics.

It is critical that we continue to monitor and assess CRISPR technology's usage and effect as it develops and its potential uses grow. We also need to have continuing ethical conversations about the appropriate use of genetic editing. By doing this, we may be able to maximize the numerous advantages of CRISPR technology while reducing the dangers and unfavorable effects of its usage.

Chapter 7:

Evolutionary Genetics

Biology's discipline of evolutionary genetics focuses on the genetic changes brought on by evolution in populations through time. The study of genetic diversity and how it develops, spreads across generations, and interacts with the environment to influence species evolution incorporates concepts from genetics, molecular biology, and evolutionary biology.

Natural selection, the process by which organisms with features that are better adapted to their environment are more likely to survive and reproduce, passing on their favorable qualities to their offspring, is one of the fundamental ideas in evolutionary genetics. These characteristics increase in population prevalence over time, causing evolutionary change.

In addition to studying mutations, genetic drift, and gene flow, evolutionary genetics also examines the genetic factors that underpin evolutionary change. Genetic drift is the random fluctuation of allele frequencies in a population brought on by chance occurrences, while mutations are changes in the genetic code that may bring new variety into a population. When people or genetic material migrate across populations, this is known as gene flow, and it may result in the mixing of genetic variety and the dissemination of advantageous characteristics.

Evolutionary genetics also aims to comprehend the links between various species and the history of life on Earth, in addition to the mechanisms behind evolutionary change. Scientists can recreate the evolutionary connections between species and comprehend how they have diverged and developed through time by comparing the DNA sequences of various creatures.

There are many real-world uses for evolutionary genetics in industries including agriculture, medicine, and conservation biology. In order to create breeding programs that are more successful, for instance, it might be helpful to understand the genetic foundation of features like disease resistance or crop productivity. Likewise, knowing the genetic variety of endangered species can assist in guiding conservation efforts to maintain genetic diversity and avert extinction.

In conclusion, the study of the genetic alterations brought about by evolution in populations through time is known as evolutionary genetics. In order to comprehend the mechanisms that underlie evolutionary change, the history of life on Earth, and the interactions between various species, it incorporates ideas from genetics, molecular biology, and evolutionary biology. grasp the intricate relationships between genetics, evolution, and the environment requires a thorough grasp of evolutionary genetics, which has implications in areas including agriculture, medicine, and conservation biology.

The Role of Genetics in Evolution

The term "evolution" refers to the changes that take place in populations of organisms through time. Genetics is a key component of this process. Genetics is used by evolutionary biologists to better understand how features are passed down across generations and change, resulting in the creation of new species and the variety of life on Earth.

Understanding that all living things have DNA, which includes the genetic information responsible for determining their qualities, is the first step in studying genetics in the context of evolution. Through reproduction, when the genetic composition of the child is formed by combining the genetic information from the parents, this information is handed down from one generation to the next.

Natural selection, which relies on genetic diversity within a population, is one of the main processes of evolution. Genetic diversity describes the variances in qualities seen among members of a group, such as disparities in height, skin tone, or illness resistance. As a result of these differences, natural selection favors people who have features that are more suited to their environment and that help them live and reproduce more successfully than those who lack such qualities. This may eventually result in the formation of new species and the accumulation of advantageous features within a population.

Natural selection depends heavily on genetics since genetic variety arises from genetics. Mutations, gene flow, and genetic drift are a few of the factors that cause genetic diversity. Mutations are haphazard alterations to DNA that may result in the creation of new alleles, or gene versions, and novel features. When people or genetic material travel across populations, this is known as gene flow. Gene flow may enhance genetic diversity or introduce novel alleles into a group. Genetic drift, which may result in the loss of genetic diversity, is the random fluctuation of allele frequencies in a population as a result of chance occurrences.

Molecular evolution is a branch of the study of genetics in evolution that makes use of genetic information to reconstruct the evolutionary connections between species and comprehend how they have diverged and developed through time. Comparing the DNA sequences of various creatures to find similarities and differences may provide information about the evolution of life on Earth.

The molecular clock, which refers to the assumption that mutations accumulate in DNA at a comparatively consistent pace across time, is one key theory in molecular evolution. Scientists can calculate how long ago different animals shared a common ancestor and how rapidly they developed since then by analyzing the amount of genetic differences between them.

Genetics offers implications in areas including health, agriculture, and conservation biology in addition to offering insights into the evolution process. In order to create breeding programs that are more successful, for instance, it might be helpful to understand the genetic foundation of features like disease resistance or crop productivity. Likewise, knowing the genetic variety of endangered species can assist in guiding conservation efforts to maintain genetic diversity and avert extinction.

The use of genetics in evolution, however, raises certain moral and societal questions. Both good and bad social and political ideas, such as eugenics, which attempted to enhance human genetic qualities via selective breeding, have been supported by research into genetics and evolution. It is critical that social responsibility and ethical values serve as the foundation for the application of genetics in evolutionary research.

The creation of novel features and the diversity of life on Earth are both made possible by genetic variety, which is a key role that genetics plays in the process of evolution. The study of genetics in evolution covers molecular evolution, which employs genetic information to reconstruct the evolutionary links between species, as well as practical applications in areas like medicine, agriculture, and conservation biology. However, moral guidelines and societal responsibilities must govern the use of genetics in evolutionary study.

Genetic adaptation is a key idea in the function of genetics in evolution. Genetic adaptation is the method through which populations develop via natural selection to become more adapted to their surroundings. This may happen either as a result of the fixation of advantageous alleles within a population or as a result of the appearance of new alleles as a result of mutation.

In response to environmental changes like climate change, the introduction of new illnesses, or the emergence of predators, genetic adaptability might be especially crucial. Populations with genetic differences that enable them to adjust to these changes more effectively are more likely to survive and procreate, which over time promotes the evolution of novel features and the creation of new species.

However, when it happens in reaction to human actions like pollution or habitat degradation, genetic adaptation may also have unfavorable effects. For instance, owing to the abuse of antibiotics, certain bacterial species have developed a resistance to them, which has resulted in the formation of illnesses that are difficult to cure because of antibiotic resistance.

The discipline of genomics, which is the study of an organism's whole genetic composition, including all of its genes and non-coding DNA, is included in the study of genetics in evolution. By enabling researchers to examine the genomes of various species and comprehend how they have developed through time, genomics has transformed the study of evolution.

Genomic research has also benefited from the development of new tools and technologies, such as the potent gene-editing tool CRISPR-Cas9, which enables researchers to make precise alterations to an organism's DNA. This has important ramifications for the study of genetics in evolution since it enables scientists to examine the role of certain genes and comprehend how they affect the development of phenotypes.

The study of genetic diversity, which refers to the variety of genetic information that occurs within and across populations, is another aspect of the function of genetics in evolution. Since genetic variety gives natural selection the starting point to work with and enables populations to adapt to changing circumstances, it is crucial for the long-term survival of species.

Overall, the topic of genetics in evolution research is intricate and dynamic, and it offers insights into the basic mechanisms underlying the development of Earth's biota. It poses significant ethical and societal issues that need to be carefully explored, but it also has significant

practical implications in areas like health,
agriculture, and conservation biology.

Human Evolution and Genetics

Genetics and human evolution are two interrelated disciplines that have aided in the scientific understanding of the genesis and evolution of our species. Genetics is the study of the transmission of genes and DNA from one generation to the next, while human evolution is the process through which people and their predecessors developed over millions of years.

An enormous area of research, involving anthropology, biology, and genetics, is the study of human evolution. Scientists have been able to trace the evolution of humans back millions of years to our earliest ancestors. Evolution is the outcome of changes in a population's genetic make-up through time.

Hominins, the oldest known progenitors of modern humans, are thought to have roamed Africa six to seven million years ago. These hominins developed into many species throughout time, each with distinctive physical traits and habits. Homo neanderthalensis, Homo erectus, and Australopithecus are a few of the best-known early human species.

The advent of Homo sapiens, or modern humans, was one of the most significant breakthroughs in the history of humanity. Around 200,000 years ago, Homo sapiens initially emerged in Africa, eventually spreading to other regions of the globe. The only human species still existing today is Homo sapiens.

Numerous elements, such as natural selection, genetic mutations, and environmental stresses, have influenced the development of humans. Natural selection happens when a population gradually starts to exhibit specific qualities that are beneficial for surviving. For instance, early hominins' ability to walk on two legs was a crucial adaptation that allowed them to cover longer distances and utilize their hands for other purposes.

In the course of human evolution, genetics has also been a key factor. The fundamental components of heredity that are carried from one generation to the next are called genes. The term "mutation" refers to changes that occur in genes that might cause the emergence of novel features in subsequent generations. While certain mutations may be detrimental, others could be advantageous and improve an organism's chances of surviving and reproducing.

The capacity to digest lactose, a milk sugar, is one instance of an advantageous genetic mutation in humans. According to estimates, this mutation first appeared in tribes that domesticated dairy animals like cows and goats roughly 10,000 years ago. These cultures could use milk as a source of sustenance thanks to their capacity for lactose digestion, which was crucial in regions with a dearth of other food sources.

Scientists have looked at the significance of epigenetics in human evolution in addition to genetic mutations. The term "epigenetics" describes changes in gene expression that do not result from modifications to the underlying DNA sequence. Environmental factors, including food, stress, and exposure to pollutants, may affect these alterations. Future generations may inherit epigenetic alterations, which may have an effect on how a species evolves through time.

The study of human genetics has also given us important new information on the evolution and history of our species. There are between 20,000 and 25,000 genes in the human genome, which is the total amount of DNA in a human cell. Height, skin tone, and susceptibility to certain malignancies are just a few of the features and illnesses that scientists have been able to link to specific genes.

The notion that every person descends from a single common ancestor who lived in Africa around 200,000 years ago is one of the most well-known findings in human genetics. This finding, which has helped shed light on the history of human migration and the development of our species, was achieved by studying DNA from various human groups throughout the globe.

The study of the hereditary causes of illness using genetics has also sparked the creation of novel cures and treatments. For instance, scientists have discovered genetic alterations linked to certain cancer types.

The Genetic Basis of Diversity

The notion that genetic variation within a species accounts for the variety of physical and behavioral qualities we see among individuals is known as the "genetic basis of diversity." This diversity results from variations in DNA sequence, which may then affect gene expression, protein function, and ultimately phenotype.

For a species to survive and adapt to changing environmental circumstances, genetic diversity is crucial. It enables the selection of beneficial features that raise an organism's chances of surviving and reproducing while also supplying the necessary building blocks for evolution.

Genetic diversity within a species is influenced by a variety of variables. The process of mutation, which modifies DNA sequences, is one of the most significant. Errors in DNA replication may cause mutations to happen spontaneously, or environmental factors like radiation or chemical exposure might cause them.

Some mutations might cause genetic problems or greater vulnerability to specific illnesses, which can be dangerous. Other mutations, however, could be advantageous and provide a benefit in certain situations. For instance, in a setting where a certain pathogen is prevalent, a mutation that provides resistance to that disease may be desirable.

Genetic recombination, or the exchange of genetic material between two separate chromosomes during sexual reproduction, is another element that leads to genetic variation. This may lead to the formation of novel gene combinations, increasing the genetic diversity of the progeny.

Another element that might increase genetic diversity is genetic drift. This happens when a population decreases due to arbitrary occurrences like natural catastrophes or population bottlenecks. In these circumstances, certain alleles could accidentally increase in frequency or decrease in frequency, changing the population's genetic composition.

The genetic foundation of diversity has significant ramifications for a number of disciplines, including biology, agriculture, and medicine. The development of customized medicine and the diagnosis and treatment of genetic problems in medicine depend on an awareness of genetic variation.

Genetic variety in agriculture is crucial for the growth of food plants with higher yields, resilience to pests and diseases, and tolerance to environmental stresses like salt or drought. To provide food security and adaptation to changing climatic circumstances, agricultural plants must maintain genetic variety.

Understanding genetic diversity is crucial for the protection of endangered species in conservation biology. A population's genetic variety enables adaptability to changing environmental factors and may act as a safeguard against the detrimental consequences of inbreeding and genetic drift.

Recent technological developments have made it possible for academics to examine the genetic underpinnings of variety in greater depth than ever before. The advent of high-throughput sequencing technology has made it feasible to sequence complete genomes, enabling the detection of millions of genetic variants inside a single person.

Additionally, these technologies have made it feasible to conduct worldwide genomic studies on populations and species. For instance, the 1000 Genomes Project seeks to sequence the genomes of 1000 people from all around the globe, creating a thorough inventory of the genetic diversity within the human population.

Advances in bioinformatics have made it feasible to examine enormous datasets and spot patterns of genetic variation in addition to sequencing methods. This has prompted the creation of new techniques for genome-wide association studies (GWAS), which seek to isolate genetic variations linked to specific characteristics or illnesses.

The genetic foundation of variety is, overall, a complicated and intricate subject with significant ramifications for several professions. For us to understand biology, medicine, and the natural world better, we must comprehend the causes that cause genetic variety as well as the functional effects of genetic variation.

Chapter 8:

Ethical and Societal Implications of Genetic Research

In recent years, there have been enormous advances in the field of genetics as a consequence of the creation of new instruments and techniques that allow the study of DNA at a level of precision that was previously unheard of. These advancements have led to the development of innovative treatments and medications that have the potential to improve the lives of millions of people across the globe and have provided new insights into the genetic underpinnings of human health and disease.

However, these advancements in genetic research also raise a number of ethical and societal concerns that need careful consideration. These effects result from the potential for genetics to reveal personal information about individuals and groups, which may have an impact on issues like social justice, privacy, and discrimination.

One of the biggest moral ramifications of genetic research is the potential for genetic discrimination. Insurance firms, employers, or other parties might potentially use genetic data for genetic profiling-based discrimination against individuals. Genetic data may reveal a lot about a person's health and disease risk. As a result, people may encounter other forms of discrimination, such as being denied access to health insurance or employment opportunities.

A further ethical conundrum is the matter of informed consent. Since such research often involves the gathering of enormous amounts of personal information, people should be informed of the potential risks and benefits of participating in it. Due to the intricacy of genetic research, people may find it difficult to fully understand the implications of their participation, and there is a danger that they may be coerced into doing so or that they may not fully understand the risks involved.

It's also crucial to consider how genetic research affects society. Genetics has the power to reveal personal information about populations, such as ancestry or a tendency for specific diseases, which might have implications for issues like social identity, collective rights, and cultural heritage. Debats over the value of diversity may also result from the discovery of genetic differences connected to traits or characteristics that society values.

Genetic research has a complex and wide range of ethical and societal repercussions. Genetic research has a lot of potential benefits, but there are also some serious drawbacks that must be carefully evaluated and addressed if genetic research is to be conducted in a manner that respects the rights and dignity of individuals and societies.

Genetic Testing and Disease Risk

Since genetic testing makes it possible to determine a person's genetic make-up and their risk of contracting certain illnesses, it has the potential to change medicine. Genetic testing involves analyzing DNA to find alterations or mutations that may be the cause of or contribute to a particular illness or condition. It may be used to make medical judgments and recommendations, diagnose or rule out a suspected genetic issue, and forecast the risk that a disease will manifest.

Through genetic testing, it is possible to find mutations linked to a higher chance of contracting conditions including breast cancer, ovarian cancer, colon cancer, Alzheimer's disease, and Parkinson's disease. People who have a family history of a certain illness or who are worried about their risk of getting a disease undergo what are known as "predictive genetic testing."

It is possible to determine the likelihood of developing an illness using a variety of genetic testing. The most popular kind is a DNA analysis, which includes obtaining a sample of cells from the person's blood, saliva, or tissue and examining the DNA for mutations linked to certain diseases. A protein analysis is a different kind of genetic test that includes examining the proteins in a person's blood or tissue for anomalies that could be linked to a disease.

The diagnosis of a genetic disorder or illness that is already present is another purpose for genetic testing. An "diagnostic genetic test" is what this is. Diagnostic genetic testing are used to support a diagnosis, pinpoint the precise genetic mutation causing the disorder, and direct management and therapy choices.

The outcomes of genetic testing may significantly affect a person's health and wellbeing. A person may be advised to have more frequent screening exams or to think about preventive measures, such as prophylactic surgery or dietary changes, if they test positive for a genetic mutation that is linked to an increased risk for a specific disease. On the other hand, if a person tests negative for a certain mutation, they may feel more at rest knowing that their risk for the illness is minimal and may not need additional screening or monitoring.

Even while genetic testing may reveal significant details about a person's health and illness risk, it's essential to remember that not all genetic tests are created equally. Genetic testing findings are not always clear-cut or simple to interpret, and some tests are more accurate and dependable than others. Additionally, it may be difficult and confusing to understand the findings of genetic tests, which calls for a high degree of knowledge.

The possibility for psychological and emotional effects with genetic testing is another crucial factor to take into account. Upon finding that they contain a genetic mutation that can be passed on to their offspring or that they have a higher risk for a certain illness, some individuals may suffer worry, dread, or despair. Genetic testing may also have an impact on a person's intimate connections, including family dynamics and reproductive decisions.

There are several current discussions and conflicts concerning the use of genetic testing for disease risk assessment, a subject that is quickly developing. One of the key issues is the possibility for genetic discrimination, which happens when someone is treated differently because of their genetic makeup, for example in the insurance or job industries. Many nations have passed legislation to safeguard people against genetic discrimination in response to this issue.

In conclusion, genetic testing is an effective method for determining a person's risk of contracting a particular illness and may be used to direct treatment choices and actions. However, the psychological and emotional effects of genetic test findings should be carefully evaluated since the interpretation of the results might be complicated. The ethical and responsible use of genetic testing is crucial as it develops in order to maximize its advantages and reduce any possible risks.

Eugenics and the Potential for Genetic Discrimination

Eugenics is the study of or belief in the possibility that genetic engineering or selective breeding might improve the genetic composition of the human species. Despite the controversial and distressing history of eugenics, modern advancements in genetic technology have renewed concerns about the prospect of genetic discrimination based on an individual's genetic make-up.

Eugenics has traditionally been associated with the forced sterilization of those deemed "unfit" or "undesirable" due to their supposed genetic traits, such as mental disease, intellectual disability, and criminality. This strategy was extensively used in the United States and Europe in the early 20th century and was commonly used as an excuse for the mass murder of millions of people during the Holocaust in Nazi Germany.

Even though the official eugenics movement has mainly died down, its ideas nevertheless influence conversations about genetic engineering and reproductive technologies. There is a particular concern that the ability to identify and manage certain genes connected to specific traits or problems may lead to bias towards persons who carry such genes.

Genetic discrimination refers to any unfair treatment of a person based on their genetic make-up, whether it relates to a job, insurance, or access to healthcare. An employer may, for example, decline to hire someone who has a gene associated with a particular illness on the grounds that doing so would raise the company's healthcare costs. In a similar vein, an insurance carrier may decline to provide coverage or increase costs for a client even though they have not yet shown the illness.

There are several laws and rules in place to avoid genetic discrimination. The Genetic Information Nondiscrimination Act (GINA) prevents employers and health insurers in the US from treating individuals differently based on their genetic information. The European Union has adopted equivalent measures in line with the General Data Protection Regulation (GDPR).

However, there are still worries about the potential for genetic discrimination in industries like life insurance, long-term care insurance, and disability insurance. Despite the fact that GINA and GDPR provide specific protections in the work and healthcare settings, there is still no consensus on how to manage this issue. They do not include every kind of insurance.

One strategy for addressing the problem of genetic discrimination is to keep genetic information private and use it only for medical purposes. For this, there would need to be rigorous limitations on who may access and use genetic information. It would also be necessary to increase knowledge and awareness of the potential detrimental impacts of genetic discrimination among the general public, insurance companies, and healthcare professionals.

Another option is to develop new insurance policies that do not assess risk using genetic information. Researchers have proposed utilizing "social insurance" models, such as those that pool risks throughout a large population and distribute rewards in line with need rather than unique risk characteristics.

Despite these challenges, it is hard to ignore the benefits of genetic testing and editing. Genetic testing could provide helpful information about a person's health and sickness risk, allowing for an earlier diagnosis and more targeted treatment. The promise of genetic engineering is the ability to treat or prevent hereditary disorders while also enhancing human talents and attributes.

It is essential to ensure that these developments are used ethically and responsibly in order to minimize the chance of harm. This necessitates ongoing debate and consideration among scientists, policymakers, and the general public about the ethical, legal, and social implications of genetic technology.

In spite of the fact that eugenics as a formal movement has mostly disappeared, its ideas continue to influence concerns about genetic technology and the potential for genetic discrimination. Even while there are laws and regulations in place to protect individuals from genetic discrimination, there are still concerns about the likelihood of bias in sectors like insurance. It is crucial to protect genetic data privacy, keep its use to a minimum for medical research, and develop unique insurance solutions that do not base risk assessment on genetic information. Overall,

The Future of Genetic Research and Society

The genetic foundation of illness is now better understood, and novel cures and treatments may be possible as a result of the enormous advancements achieved in genetic research in recent years. The use and regulation of genetic research, however, poses significant ethical, social, and legal issues as it develops.

The growing use of genetic testing to determine the genetic basis of illnesses and disorders is one of the most important advancements in genetic research. Genetic testing may help to pinpoint certain genetic variants that put a person at higher risk for conditions including breast cancer, Alzheimer's disease, and cystic fibrosis. People may utilize this knowledge to make educated choices about their future and health, as well as to design more individualized and focused therapies.

Genetic testing, however, also brings up issues with privacy and the possibility of discrimination based on a person's genetic make-up. There is a chance that people might be subject to discrimination based on their genetic makeup in settings including jobs, insurance, and healthcare. Laws and regulations have been created to protect people from genetic discrimination in order to address these issues, such as the Genetic Information Nondiscrimination Act (GINA) in the United States.

The use of gene editing tools like CRISPR-Cas9 is a significant advancement in genetic research. By precisely targeting and altering certain genes, these technologies enable researchers to develop novel therapies for genetic illnesses and disorders. Nevertheless, the use of gene editing poses serious moral concerns around the possibility of "designer babies" and the establishment of genetic "superiority." Inequality and prejudice in society might result from the use of gene editing to produce people with certain features or attributes.

In addition to these concerns, genetic research also presents issues with the ownership and management of genetic information. Who owns the rights to a person's genetic data, and how ought that data to be used and distributed? In the context of large-scale genetic research programs that gather and analyze genomic data from hundreds or even millions of people, these problems assume particular significance.

Despite these obstacles, genetic science has a bright future in terms of enhancing human health and wellbeing. Genetic study may result in the creation of novel cures and treatments, as well as a better understanding of the origins of human variation and human evolution.

The social, ethical, and legal ramifications of genetic research should be continuously discussed and debated, however, to guarantee that it is utilized ethically and responsibly. A wide variety of stakeholders, including scientists, politicians, healthcare professionals, and the general public, must be involved in this.

Establishing precise rules and regulations for the use of genetic research, including rules for the use of gene editing technologies and rules for the use and sharing of genetic data, is one possible approach. These rules might make it easier to preserve people's rights and privacy while doing genetic research in a moral and responsible way.

Increased public education and knowledge of genetic research's possible advantages and disadvantages is another option. This may include expanding access to genetic counseling and education as well as raising public awareness of scientific issues.

Overall, genetic science has a bright future in terms of enhancing human health and wellbeing. However, it is crucial to address the social, ethical, and legal consequences of genetic research and to include a wide variety of stakeholders in continuing discussion and debate in order to guarantee that it is utilized ethically and responsibly. By doing this, we may increase the potential gains from genetic research while lowering its dangers and hazards.

Chapter 9:

Conclusion

The Future of Genetics

One of the most fascinating and enticing fields of scientific study is the future of genetics. We are learning more and more about the function of genes in human health and illness as our grasp of the genetic code deepens. The possible applications of genetic research are many and varied, ranging from gene editing to customized treatment. We shall examine some of the most promising fields of genetic research in this article, along with any potential repercussions for the practice of medicine and society at large.

Personalized medicine is one of the most exciting fields of genetic research. The concept of personalized medicine holds that a patient's distinct genetic profile may inform their medical care. This implies that physicians may use a patient's genetic information to determine which treatments are most likely to be helpful and which could have negative side effects on their medical care. This implies that physicians may use a patient's genetic information to determine which treatments are most likely to be helpful and which could have negative side effects. Additionally, it may be used to forecast a person's likelihood of contracting certain illnesses, such as cancer or heart disease, and to create specialized preventive measures. The ability to tailor treatment and make it more effective, efficient, and sensitive to the requirements of each patient has the potential to change healthcare.

Gene editing is a new field of genetic study with huge promise. Gene editing is the process of altering a person's DNA to fix genetic flaws or improve certain qualities. By removing the faulty genes that cause genetic disorders like cystic fibrosis or Huntington's disease, this technique holds the potential to treat these problems. Additionally, it might be utilized to improve abilities like intellect or athleticism. Although there is ongoing discussion on the morality of gene editing, it is obvious that this technology has the power to revolutionize how we see human biology and genetics.

Numerous other fields of genetics research are also showing considerable promise, including customized medicine and gene editing. For instance, scientists are utilizing genetic data to comprehend the function of genes in complicated disorders like diabetes, Alzheimer's disease, and cancer. It may be possible to utilize this knowledge to create fresh therapies and preventative measures for certain ailments.

Epigenetics is another field of genetic study that is causing a lot of enthusiasm. The term "epigenetics" describes modifications to gene expression that do not result from alterations in the DNA sequence itself. Instead, chemical alterations to the DNA molecule, which may affect how genes are read and translated into proteins, are to blame for these changes. It is obvious that epigenetic changes play a crucial role in many areas of human health and illness, despite the fact that researchers are just now starting to comprehend the intricate interactions between genetics and epigenetics.

Technology advancements are also expected to have an impact on the future of genetics. For instance, the ability to sequence the complete human genome in a matter of days thanks to the advent of quicker, more affordable DNA sequencing technology has already changed genetics research. Doctors will be able to utilize genetic data in standard medical practice as this technology advances and becomes more generally accessible and reasonably priced.

Technology advancements outside of DNA sequencing are also anticipated to have a significant influence on genetics research. For instance, the advancement of machine learning and artificial intelligence algorithms may make it simpler to evaluate vast volumes of genetic data and spot patterns and links that may not be immediately apparent to human researchers. The genetic foundation of illness may become more clear as a result, which may aid in the development of more potent therapies and preventative measures.

Finally, it's critical to think about how genetics research will affect society as a whole. Despite the immense potential advantages of this study, it is crucial to think about the moral and societal ramifications of utilizing genetic information to alter human characteristics or make medical choices. For instance, using gene editing to improve specific qualities can cause the wealth gap to widen since only those who can afford these therapies will have access to them.

Final Thoughts and Reflections

The book that covers the aforementioned topics offers an engrossing glimpse into the world of genetics. It addresses the fundamental concepts and theories of genetics while demonstrating how important genetics is to modern society, agriculture, and medicine. I appreciate the author's use of clarity and depth to explain complex scientific concepts in a manner that is clear to laypeople as I analyze the book's content.

Chapter 1 does an outstanding job of introducing the field of genetics. It emphasizes the historical significance of genetics and how it influenced modern science. The chapter also examines the contributions of significant figures in genetics research, including Gregor Mendel, Francis Crick, and Thomas Hunt Morgan. In order for readers to know what to expect in the next chapters, the author provides a brief summary of the book.

Chapter 2 examines the principles of heredity, starting with Mendelian genetics. The author describes Mendel's theorized laws of inheritance, the cornerstone of modern genetics. Following is an explanation of meiosis and chromosomal segregation, which are crucial for preserving genetic diversity. Along with sex-related characteristics, this chapter also discusses how men and women inherit certain traits differently.

Chapter 3 discusses the composition and function of DNA, the molecule that contains genetic information. The research looks at the replication, transcription, and DNA structure of DNA, as well as how Watson and Crick discovered DNA. The chapter sets the stage for comprehension of the molecular basis of genetics and gets the reader ready for the subsequent chapters.

We examine genetic alterations and how they affect health in Chapter 4. There includes a full treatment of the many types of genetic mutations, including chromosomal, frameshift, and point mutations. Additionally included in this chapter are inherited conditions and illnesses such Huntington's disease, sickle cell anemia, and cystic fibrosis. Gene therapy, which has the potential to treat inherited diseases, is also discussed.

Chapter 5 addresses the disciplines of genetics and personalized medicine. The Human Genome Project, which was finished in 2003, is discussed at the chapter's beginning. Following that, the author goes into great length on how genome analysis and sequencing have changed our understanding of genetics. The possibility of personalized medicine, which is medication made especially for a person based on their genetic makeup, is also discussed.

Chapter 6 focuses on biotechnology and genetic engineering. The author investigates genetic engineering in agriculture and animal breeding, highlighting both its benefits and potential drawbacks. In his examination of the many ethical concerns surrounding genetic engineering, the author discusses the ethics of this field. The chapter is concluded with a description of CRISPR, a cutting-edge gene-editing method with the potential to cure genetic diseases.

Chapter 7 examines the analysis of the genetic bases of evolution. By studying how mutations, gene flow, and genetic drift may influence a population's genetic composition, the author highlights the significance of genetics in evolution. The author also looks at the genetic evidence for the history of human evolution in his discussion on human evolution and genetics.

Chapter 8 talks about how genetic science affects society and ethics. The chapter begins with a discussion of genetic testing as a means of assessing the likelihood of contracting a disease. It also considers the likelihood of hereditary prejudice and eugenics. The author presents a balanced approach, weighing the benefits and drawbacks of genetic research while highlighting the importance of ethical concerns in this field.

Chapter 9 of the book discusses the future of genetics and brings the book to a close. While underlining how genetics has the potential to transform society, health, and agriculture, the author also highlights some potential risks. The need of ethical and responsible research is emphasized as the chapter comes to a close to ensure that genetics is used for the good of humanity.

The book offers a complete analysis of the field of genetics, highlighting its significance in many areas of life and outlining its fundamental ideas. Due to the author's excellent job of presenting complex scientific concepts in a form that is intelligible to the public, the book is appropriate for a wide range of readers.

One of the book's main strengths is the careful exploration of the potential benefits and drawbacks of genetic research. Examining the moral concerns raised by genetic research, the author highlights the need of morally responsible research methods for ensuring that genetics is used for the good of humanity.

Another feature of the book's strength is the author's emphasis on how genetics has the potential to transform a variety of sectors, including health, agriculture, and society. The book has a strong emphasis on how genetics has progressed biotechnology, genetic engineering, and personalized medicine, giving readers a glimpse into how this field may shape the future.

Overall, the book provides a detailed explanation of the fundamental concepts, theories, and applications of genetics and is an excellent primer on the field. As I evaluate the content of the book, I appreciate the author's efforts to make it intelligible to a range of readers while still providing a comprehensive analysis of the subject. Since it offers a solid foundation for further research into this interesting field, everyone with an interest in genetics should read the book.

As this book on genetics draws to a close, the huge potential for this field to fundamentally transform many aspects of our life hits me. Future developments in genetics, such as genetic engineering and personalized medicine, might have an impact on how humanity progresses.

However, as the book has made apparent, this option creates important ethical issues. Given the broad implications of genetic research, it is essential to carry out morally responsible research and use genetics for the good of humanity.

The book also illustrated the complexity and dynamic nature of genetics, which is a field that is always undergoing fresh research. As a consequence, it's essential to stay up to date on the most current developments in the field and to be receptive to new ideas.

When I think about the book's subject matter, the importance of genetics in fostering variety and improving knowledge of our shared humanity hits me. Genetics has shown our common humanity, and our unique differences are proof of the enormous diversity of the human condition.

Last but not least, this book on genetics serves as a fantastic introduction to the field, providing a detailed examination of its underlying concepts, ideas, and applications. The book addressed the need of ethical considerations and responsible research, as well as how genetics has the ability to change numerous businesses.

It is certain that as time goes on, genetics will have a greater and greater impact on our environment. Therefore, it is essential to continue studying and learning more about this interesting subject, as well as to work together to ensure that genetics is used for the good of all humans.

Printed in Great Britain
by Amazon

29107133R00086